International Sport Coaching Framework

VERSION 1.2

International Council for Coaching Excellence

Association of Summer Olympic International Federations

Leeds Metropolitan University

Copyright © 2013 by the International Council for Coaching Excellence, the Association of Summer Olympic International Federations, and Leeds Metropolitan University

Published by Human Kinetics, Champaign, Illinois, United States

ISBN-10: 1-4504-7127-7 (print)
ISBN-13: 978-1-4504-7127-5 (print)

Contents

Acknowledgements   4

Introduction: A Step Forward for Coaching   5

1   Coaching Today   6
Ever-Higher Expectations
An Athlete-Centred Orientation
A Stronger Development Network

2   Coaching Framework Foundations   9
Practical and Flexible Applications
Global Relevance

3   Coaching in Context   13
Sport-Specific Emphasis
Volunteer or Paid Status
Primary Functions
Key Responsibilities

4   Coaching Focus   19
Values
Sport Participation
Athlete Development
Contextual Fit

5   Coaching Roles   24
Role Requirements
Staff Assignments and Synergy

6 Coaching Knowledge and Competence — 29
Knowledge Areas
- Professional Knowledge
- Interpersonal Knowledge
- Intrapersonal Knowledge

Competences
- Functional Competence
- Task-Related Competence

7 Coaching Objectives — 34
- Developing the Whole Athlete
- Teaching Lifelong Lessons

8 Coach Development — 37
- Long-Term Process
- Educational Curriculum
- Experiential Learning and Mentorship
- Delivery by Coach Developers

9 Coach Certification and Recognition — 43
- Educational Requirements
- Qualifying Standards
- Awards and Designations

10 Coaching Framework Applications — 48
- Create High-Quality Coach Education and Development Programmes
- Evaluate and Improve Existing Programmes
- Define Areas for Research and Evaluation
- Consider and Make Political Decisions
- Stimulate Global Exchange
- Promote Further Refinement

Glossary — 53

About the Authors — 54

Acknowledgements

The International Council for Coaching Excellence (ICCE) and the Association of Summer Olympic International Federations (ASOIF) wish to acknowledge a range of individuals and organisations for the production of this document. The *International Sport Coaching Framework Version 1.2* reflects a process of consultation, development and collaborative research. This process has involved ICCE, ASOIF, the World Anti-Doping Agency (WADA), Olympic Solidarity and their respective members. In addition, Leeds Metropolitan University (LMU) is now the home of the ICCE Global Coaching Office and has contributed extensively to the development of the document through its staff and research programmes in sport coaching and coach development.

Contributors to consultation meetings in Köln (April 2011), Paris (September 2011), Madrid (February 2012), Beijing (April 2012), Sofia (May 2012), Leeds (March 2013) and during the ASOIF development seminar in Lausanne (April 2013) have ensured that the process of developing the document has had a wide international reach. The editors of the publication, Pat Duffy, Mark Harrington and Sergio Lara Bercial, have done excellent work capturing the input from the three-year developmental process.

We wish to thank the members of the Joint ICCE-ASOIF Working Group for their efforts:

Joint chairs: Pat Duffy (LMU and ICCE) and Marisol Casado (ASOIF)

ICCE Working Group members: José Curado, Frank Dick, Lutz Nordmann, Frederic Sadys, Desiree Vardhan and Bingshu Zhong

ASOIF: Andrew Ryan, Jacqueline Braissant, Miguel Crespo, Kelly Fairweather (and deputised by Tayyab Ikram), Ivo Ferriani (and deputised by Georg Werth), Elio Locatelli and the members of the ASOIF Education and Development Group, chaired by Mark Harrington

International Olympic Committee Entourage Commission: Clive Woodward

Olympic Solidarity: Yassine Yousfi

World Anti-Doping Agency: Rob Koehler (and Lea Cleret)

UK Sport: Ollie Dudfield and Priya Samuel

ICCE: John Bales, Adrian Burgi, Penny Crisfield (lead author of chapter 8), Sergio Lara Bercial (technical officer of the project), Karen Livingstone and Ladislav Petrovic

International federations (technical support to the process): Dan Jaspers, Kyle Philpotts, Rosie Mayglothling and John Mills

Research: Jean Côté, Wade Gilbert, Cliff Mallett, Julian North and Pierre Trudel

European Commission: Bart Ooijen

We also wish to thank ICCE lead partner, South African Sports Confederation and Olympic Committee (SASCOC), for piloting key elements of the *Framework*. And this publication is a result of the ICCE's partnership with Human Kinetics in support of coaching and coach education throughout the world. We are particularly indebted to HK's Ted Miller for his skilled guidance and input.

Introduction

A Step Forward for Coaching

Version 1.1 of the *International Sport Coaching Framework (ISCF or Framework)* stimulated much interest and garnered an encouraging amount of support after its debut as a consultation document at the 2012 Summer Olympics and Paralympics in London. That document provided sport federations, coaching organisations, international federations and educational institutions with key considerations to support the design, benchmarking and refinement of their coach education and development programmes.

During the past year ICCE and ASOIF have intensified the research and consultation process. In this effort, we benefitted from a wealth of feedback and findings from around the world. Reviews of Version 1.1 were mainly positive, reinforcing the potential value of such a resource. Constructive critiques pointed to the need to make the language more accessible, prioritize key concepts and provide examples of application from the field. Others expressed concern that the *Framework* might be binding, having perceived the document as a set of proposed mandates.

Version 1.2 responds to that feedback with less formal language, a new glossary, greater emphasis on topics of greatest pertinence and highlights of good practices and valuable research that will be useful to a range of stakeholders. Furthermore, we make the point throughout the document that multiple effective approaches need to be tailored to sport- and country-specific circumstances.

Version 1.2 of *ISCF* is, therefore, both an authoritative and flexible reference document that facilitates the development, recognition and certification of coaches. We hope you will find it a useful guide, one that you can apply fully or in part to suit your own context and to evaluate and report what you have found. Though we believe this document is a big step forward, we will continue to research and evaluate each element of the *Framework* and consider all feedback as we seek to refine it in forthcoming years. Version 2.1 is planned for release at the Global Coaches House in Rio de Janeiro in 2016.

John Bales
President, ICCE

Andrew Ryan
Director, ASOIF

Professor Andrew Slade
Deputy Vice Chancellor, LMU

Coaching Today

Coaches play a central role in promoting sport participation and enhancing the performance of athletes and teams. In accordance with internationally recognized and domestic sporting codes, coaches guide children, players and athletes. In nearly 200 countries millions of volunteer and paid part-time and full-time coaches deliver sporting opportunities to millions of participants.[1]

Ever-Higher Expectations

In addition to their core role, coaches contribute to the development of athletes as people, teams as cohesive units and communities with a shared interest. Coaching also can contribute to social aims by promoting activity and health; coalescing citizens behind a common entity; and generating economic activity through employment, education, purchase of equipment, use of facilities and attendance at events.

Coaching is in its most dynamic era in history. Coaches work with increasingly diverse populations and face heightening demands from their athletes, their athletes' parents, administrators and fans. Coaches are required to fulfill a variety of roles that may include educator, guide, sport psychologist and business manager. At higher levels of competition coaches are asked to emphasize positive interaction and overall development of athletes rather than simply the win–loss record.[2] There is greater accessibility to information and visibility to a larger community in the digital age. All of these factors make coaching more exciting and taxing than ever before.

Coaches, therefore, represent a vast positive resource to activate and mobilize children and athletes in a variety of sports. Much of this effort is fuelled by volunteer effort and the market economy. Governments, federations and other organisations invest in coaching to varying degrees, and this document will further enhance the nature and focus of such investment.

An Athlete-Centred Orientation

Coaches have a responsibility to improve and expand their capabilities on an ongoing basis to fully meet the needs of the athletes they serve. The organisations that employ them owe it to coaches to ensure they have sufficient educational footing, philosophical orientation and resources to fulfil the duties expected of them.

Coaches who are supported in this way will become even better equipped to adopt an athlete-centred focus. In support of their huge contribution to sport delivery, coaches should enhance their expertise and effectiveness in the achievement of agreed outcomes with their athletes.[3]

Athlete-centred coaching, therefore, suggests a commitment to lifelong learning by the coaches themselves. It also highlights the need for federations and coaching organisations to clearly map out the stages and competences associated with the long-term development of coaches.[4]

> Sport coaches play a central role in the development of children and athletes at all levels.

A Stronger Development Network

Governmental bodies and sport organisations have invested in programmes to enhance the quality of coaching and increase the number of coaches to meet present and future needs. The International Olympic Committee (IOC) has recognized the significance of coaching through its Olympic Solidarity (OS) programmes and the recent creation of the Entourage Commission.

National and international federations play a lead role in developing coaches through established sport-specific codes and delivery networks. However, implementation and funding of coach development can be challenging. Fortunately, in many cases, a wider network of educational institutions, public authorities, businesses, organisations and individuals also provide invaluable resources and serve as key partners in this critical endeavour. Figure 1.1 outlines the network of organisations involved in coach development, a jigsaw that will vary according to the structures in each sport and country.

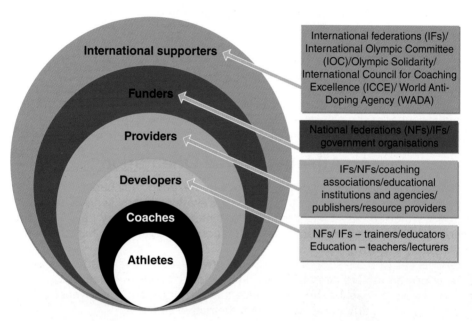

FIGURE 1.1 Network of organisations involved in coach development.

---

[1] In the UK research has found that 1.1 million coaches deliver sport to over 10 million participants every year (North, J. 2009). *The UK coaching workforce*. Leeds: Sports Coach UK. The sport of football alone has as many as 265 million players worldwide (FIFA 2013) (www.fifa.com/worldfootball/bigcount/index.html downloaded 3 July 2013).

[2] This theme has been taken up in the recent publication: European Commission Sport Unit (2012). *EU guidelines on dual careers of athletes*. Brussels: European Commission.

[3] Coaching effectiveness and expertise were defined by Côté, J., and Gilbert, W. (2009). An integrative definition of coaching effectiveness and expertise. *International Journal of Sports Science and Coaching*, 4(3): 307–323. These issues are addressed further in chapter 6.

[4] See South African Sports Confederation and Olympic Committee. (2012). *The South African model for long-term coach development*. Johannesburg: Author.

Coaching Framework Foundations

The welfare of athletes is the foremost concern to coaches in designing, implementing and evaluating appropriate practices and competitions. While many commonalities exist throughout the global coaching community, unique characteristics in coaching prevail in every sport and country. National and international sport federations define and regulate the more universal coaching codes. National and

local organisations implement more customised coaching programmes and guidelines to address particular objectives and issues.

Sport is best served when the principles and policies of the international and national federations dovetail with national programmes and the needs and experiences of coaches and athletes in local communities. For example, in the respective developments of the *South African Coaching Framework* and the *UK Coaching Framework*, there has been increasing alignment between national coach education programmes and those of the international federations.[1] This has resulted in better experiences for the coaches, increased mobility and, in some cases, enhanced funding at the national level. It is also the case that internationally funded programmes such as those provided by OS are increasingly seeking sustainability of coach education programmes in line with national and local needs.

The growing appreciation of coaching and the challenges that accompany the role have highlighted the need for a more common language and set of criteria to inform the development and qualification of coaches. Given the advances in communication technologies and convenient means of travel, many of the obstacles that may have once blocked the flow of knowledge, information and research about coaching have been addressed to varying degrees in various sports and nations. Thus, there is an opportunity to create a vibrant global dialogue and professional language among organisations interested in developing coaches in a more systematic and sustained manner. The *International Sport Coaching Framework (ISCF or Framework)* provides common principles, concepts and tools that can be applied to the needs of various sports and countries. The recognition of coaching qualifications, nationally and internationally, as well as the mobility of coaches will be enhanced.

The *ISCF* can assist in coach development, recognition and certification. The core role of national and international federations in defining and leading sport-specific coach development is recognized as well as the different and autonomous national systems for coach education and certification.

> **The *International Sport Coaching Framework* is an internationally recognised reference point for the development of coaches. It is responsive to the needs of different sports, countries, organisations and institutions and provides benchmarks for the recognition and certification of coaches.**

By establishing key terms and methodologies for application, the *ISCF* provides a benchmarking tool to assist each sport and each country in refining existing programmes and planning new programmes. It will also set the stage for increased synergy between national and international organisations in support of coach education and development.

Practical and Flexible Applications

A central feature of the *ISCF* is the recognition that coaching occurs through diverse sporting codes, each with unique rules, structures and traditions. Circumstances will vary depending on the stage of development of the systems in various sports and countries. In all cases, it should be informed by front-line need as depicted in figure 2.1.

Through this process, the complementary roles of international and national federations and government-based organisations will be more clearly defined, enhancing the reach of effective coach education and development programmes across sports and nations. The recognition of coaching qualifications and coach mobility will also be enhanced.

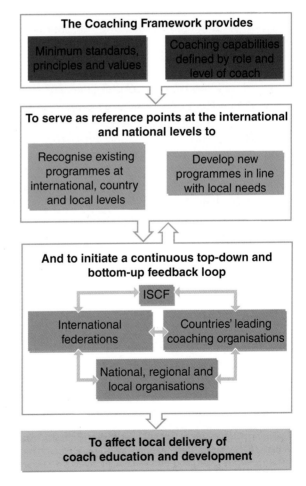

FIGURE 2.1 Application of the *International Sport Coaching Framework*.

Global Relevance

Principles, concepts and tools guide the support and development of coaches at all levels:

- **Terminology.** Key terms and meanings are provided, enhancing the possibility of collaboration for the benefit of coaches and coaching. A common language, whilst recognising linguistic and cultural differences, also opens up greater possibilities for programme design, the recognition of coaching qualifications and the mobility of coaches. The terminology is informed by recent research on coaching expertise (such as Côté and Gilbert, 2009;[2] Côté et al., 2007[3]) and by recent publications at national and international levels.[4]

- **Role definition.** The *Framework* defines the roles taken on by coaches according to levels of competence and responsibility and the populations they serve (coaching domains) in **participation sport** and **performance sport**. Role definition also addresses the status of coaches based on full-time paid, part-time paid and volunteer involvement. Codes of conduct in meeting those responsibilities are also addressed.

- **Coaching competence and standards.** The *ISCF* assists in mapping coaching competences with coaching roles; defining minimum standards for training, certifying and evaluating coaches; and enhancing the effectiveness of coaching in various contexts. International federations have begun to apply such processes of mapping, including the International Rugby Board, International Tennis Federation, International Sailing Federation, Badminton World Federation, International Table Tennis Federation, and International Association of Athletic Federations.[5] National **coaching systems** have also begun to use the *Framework*. For example, the *South African Model for Long-Term Coach Development* was drafted using Version 1.1 as a key point of reference.

---

[1] South African Sports Confederation and Olympic Committee. (2011). *The South African Coaching Framework*. Johannesburg: Author. Sports Coach UK. (2008). *The UK Coaching Framework*. Leeds: Coachwise.

[2] Côté, J., & Gilbert, W. (2009). An integrative definition of coaching effectiveness and expertise. *International Journal of Sports Science and Coaching*, 4(3): 307–323.

[3] Côté, J., Young, B, North, J., & Duffy, P. (2007). Towards a definition of excellence in sport coaching. *International Journal of Coaching Science*, 1(1), 3-16.

[4] European Coaching Council. (2007). *Review of the EU 5-level structure for the recognition of coaching qualifications*. Koln: European Network of Sport Science, Education and Employment. In addition, the various coach education and development publications of Badminton World Federation, Coaching Association of Canada, Federation Equestre Internationale, International Association of Athletic Federations, International Table Tennis Federation, International Tennis Federation, International Rugby Board and others have informed the research and development process as well as those publications outlined in footnotes 1 to 3.

[5] In the case of IAAF, a formal mapping process appears in Duffy, P., Petrovic, L., & Crespo, M. (2010). The European framework for the recognition of coaching competence and qualifications: Implications for the sport of athletics in Europe: A report to European Athletics. *New Studies in Athletics*, 26(1), 27-41.

3

Coaching in Context

Coaching is a relational, not isolated, activity. Coaches must understand, interact with and influence the settings in which they work. Coaches should, therefore, build functional relationships with athletes and the **entourage** while seeking to implement effective and ethical practice and competition programmes. Coaches' voices should also be heard in organisational decisions, especially those affecting the development of their athletes.[1]

Central to coaching in all contexts is the creation of practice and competition opportunities that result in desired outcomes for athletes. At the core of a coach's role is guiding the improvement of athletes in sport-specific contexts, taking account of athletes' goals, needs and stages of development.[2]

To execute these duties proficiently, coaches must be suitably informed, active, skilled and qualified. Coaching effectiveness is gauged by the consistency with which positive outcomes for athletes and teams are achieved, reflected only in part by competitive success. Indeed, a coach who unifies a group for a common

> Coaching is a process of guided improvement and development in a single sport and at identifiable stages of development.

purpose or provides skills for lifelong participation is every bit as successful as the league title-winning coach.

The challenge to maximize effectiveness with various groups of athletes and changing circumstances is part of the allure and richness of coaching. Because of the diversity of the role and contexts, the delivery of coaching and what is deemed successful will always be situation specific.

Sport-Specific Emphasis

The sport-specific nature of coaching is central to the identity of the coach. National federations regulate and structure sport in individual countries just as international federations govern sport on a global basis. Coaches, therefore, work with rules and laws (game formats and competition structures) that are inextricably linked to their chosen sports.

National, regional and local authorities deploy coaches to achieve objectives in participation and performance. In turn, coaches in schools, clubs and communities focus on what they and their athletes wish to achieve. Thus, coaches should develop programmes to meet the needs of their athletes *and* contribute to the goals of their organisations (see figure 3.1).

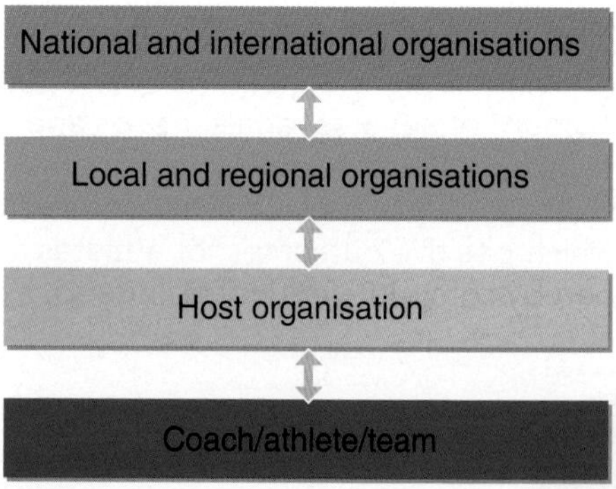

National and international organisations
↕
Local and regional organisations
↕
Host organisation
↕
Coach/athlete/team

FIGURE 3.1 Layers of coaching engagement.
Adapted from Bronfenbrenner[3] and Carlson.[4]

Volunteer or Paid Status

Research data from various countries show that the coaching community (i.e., **coaching status**) consists of volunteer, part-time paid and full-time paid coaches as outlined in figure 3.2.[5]

The ratio of these categories varies according to sport and context, leading to variations in how the identity of coaches is seen (see table 3.1[6]). Coaching differs from other professions due to its high proportion of volunteers, many of whom are parents and older athletes who take on immediate coaching needs. For this reason coaching is seen as a blended professional area. In many cases, volunteers undertake **pre-coaching roles** where a formal decision to coach or gain qualifications has not yet been made.

Sport also benefits from more experienced coaches. These coaching experts serve many roles, but perhaps none more important than supporting and managing the work of other coaches, including volunteer coaches and pre-coaches.

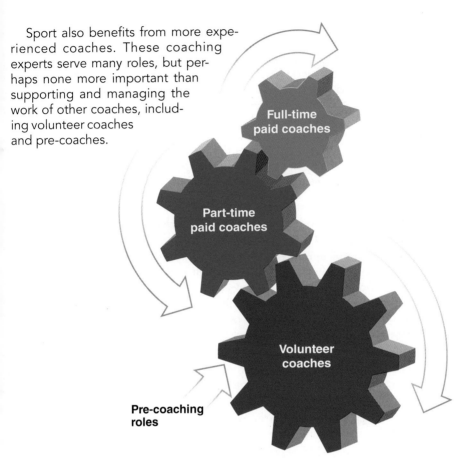

FIGURE 3.2 Categories of coaching status.

TABLE 3.1 **Coaching Status and Identity**

Professional identity	Blended identity	Voluntary service identity
High proportion of paid coaches committed to professional standards, organisation and professionalisation.	Majority of unpaid coaches with a substantial number of paid roles. Commitment to a process of professionalisation and organisation.	High proportion of volunteer coaches with few, if any, paid roles. Commitment to service and volunteering for its own sake. Varied agenda on professionalisation and organisation.

⟵ Sport, country and organisational variation ⟶

Primary Functions

The primary functions of coaches have been extensively researched in recent years.[7] Several functional and competence-based frameworks have been developed at national and international levels, such as those of some international federations, the European Coaching Council, Zone VI in Africa, SASCOC in South Africa and the National Coach Certification Programme in Canada. The *International Sport Coaching Framework* specifies six primary functions, all helping to fulfil the core purpose of guiding improvement and development. These primary functions have been derived from consultation, a review of the literature and primary research.[8]

1. **Set the vision and strategy.** The coach creates a vision and a strategy based on the needs and stages of development of the athletes and the organisational and social context of the programme.
2. **Shape the environment.** The coach recruits and contracts to work with a group of athletes and takes responsibility for setting out plans for specified periods. The coach also seeks to maximize the environment in which the programme occurs through personnel, facilities, resources, working practices and the management of other coaches and support personnel.
3. **Build relationships.** The coach builds positive relationships with athletes and others associated with the programme, including personnel at the club, school, federation and other levels. The coach is responsible for engaging in, contributing to and influencing the organisational context through the creation of respectful working relationships.
4. **Conduct practices and prepare for and manage competitions.** The coach organises suitable and challenging practices using effective techniques (e.g., practice design, demonstration, observation, feedback) to promote learning and improvement. The coach prepares for targeted competitions and also oversees and manages the athletes in these competitions.
5. **Read and react to the field.** The coach observes and responds to events appropriately, including all on- and off-field matters. Effective decision making is essential to fulfilling this function and should be developed in all stages of coach development.[9]
6. **Learn and reflect.** The coach evaluates the programme as a whole as well as each practice and competition. Evaluation and reflection underpin a process of ongoing learning and professional development.[10] The coach also supports efforts to educate and develop other coaches.

> **Coaches perform their role in social and organisational environments, varying in nature between sports and contexts.**

These primary functions describe how coaches accomplish their aims in general terms. Substantial variation may exist depending on the nature of specific **coaching roles** and circumstances. Also, experienced coaches typically are more engaged than early-stage coaches in all of the functions. However, all coaches should strive to fulfil these primary functions regardless of experience. The knowledge and competence associated with these functions are examined in detail in chapter 6.

The primary functions are interrelated and occur in a cyclical process of improvement that includes planning, implementation, review and adjustment as outlined in figure 3.3.

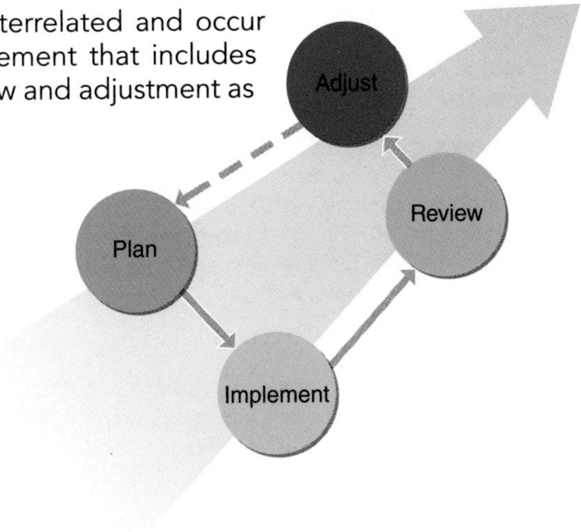

FIGURE 3.3 The cycle of coaching and continuous improvement.

The process also recognises that coaches operate in cycles ranging in duration from just one practice session to a portion of a season, an entire season, a quadrennium, or the major part of an athlete's career.

Key Responsibilities

Coaching functions focus on enhancing athlete performance and personal development. The premise of an athlete-centred approach is the protection of and respect for the integrity and individuality of those with whom coaches work. Coaches have a particular responsibility to safeguard and protect children and young people in their care and recent international work has focused on this responsibility.[11]

Coaches must also abide by the international and national rules relating to anti-doping as defined by WADA in the CoachTrue programme (coachtrue.wada-ama.org). Work is on-going between ICCE and WADA on the integration of anti-doping knowledge and competence into the education of coaches. The clear expectation is that coaches will perform their duties in an ethically responsible way, play by the rules at all times and protect the integrity of sport.

[1] Duffy, P., North, J., & Curado, J. (in press). *CoachNet: a study to identify the current and potential future 'voice of the coach' within the European Union*. Report to the European Union Commission. Leeds: Leeds Metropolitan University.

[2] This definition builds on the earlier work of European Coaching Council (2007), *Review of the EU 5-level structure for the recognition of coaching qualifications*. Koln: European Network of Sport Science, Education and Employment.

[3] Bronfenbrenner, U. (1979). *The ecology of human development*. Cambridge, MA: Harvard University Press.

[4] Carlson, R. (1993). The path to the national level in sports in Sweden. *Scandinavian Journal of Medicine and Science in Sports*, 3, 170-177.

[5] Research evidence is drawn from North, J. (2009). *The UK coaching workforce*. Leeds: Sports Coach UK; Duffy, P., Hartley, H., Bales, J., Crespo, M., Dick, F., Vardhan, D., Normann, L., & Curado, J. (2011). Sport coaching as a 'profession': Challenges and future directions. *International Journal of Coaching Science, 5(2)*, 93-124. The conceptual framework for coaching status has been adapted from South African Sports Confederation and Olympic Committee. (2011). *The South African Coaching Framework*. Johannesburg: Author.

[6] Adapted from Duffy et al. (2011). See note 5.

[7] These have included Abraham, A., Collins, D., & Martindale, R. (2006). The coaching schematic: Validation through expert coach consensus. *Journal of Sports Sciences*, 24(6): 549–564; Côté, J., & Gilbert, W. (2009). An integrative definition of coaching effectiveness and expertise. *International Journal of Sports Science and Coaching*, 4(3): 307–323; and others.

[8] This research has been conducted by Leeds Metropolitan University (LMU) in partnership with ICCE and has included a literature review, a study on the perspectives of university coaches and researchers on the primary functions and a study on 'serial winning' coaches around the globe. Each of these studies is being prepared for publication, and full references will be posted on the LMU and ICCE websites when completed.

[9] Abraham, A., & Collins, D. (2011). Taking the next step: Ways forward for coaching. *Quest* (63), 366-384. This publication outlines the centrality of contextual decision making in the role of the coach.

[10] See Werthner, P., & Trudel, P. (2006). A new theoretical perspective for understanding how coaches learn to coach. *The Sport Psychologist*, 20(2): 198–212.

[11] The Safeguarding Children in Sport Working Group is an international initiative, working to make sport safer for children. One key element of this work is the development of a set of international standards for safeguarding which are currently being piloted around the globe. These standards will help organisations develop the systems they need in place to make sport safer. Find out more at: http://assets.sportanddev.org/downloads/international_sports_safeguarding_children_standards_draft.pdf

4

Coaching Focus

Coaches face many distractions and pressures that can affect their vision and role. By focusing their attention and energies, they will more effectively meet the needs of athletes and optimize their performance in their specific sports and national contexts.

Coaching Focus

Values

Coaches, like everyone, direct their attention and actions towards the things they value. A prerequisite of coaching should be a strong interest and commitment to the positive sport experience and development of each athlete. Coaches should develop an ethically grounded coaching philosophy, supported by objectives that are athlete focused and take account of the context in which coaching occurs.

To guide improvement on a sustained basis, coaches must be attuned to the needs and progress of the athletes. In turn, **coach development** programmes should enhance the competences and knowledge required for working with specific categories of athletes.

Sport Participation

Research and evidence from the field,[1] have identified two types of engagement in sport: participation sport and performance sport. The former emphasizes involvement and enjoyment; the latter accentuates competition and achievement.

Within each of these two types of sport engagements are three subdivisions:

- Participation
  1. Children
  2. Adolescents
  3. Adults

- Performance
  1. Emerging athletes
  2. Performance athletes
  3. High-performance athletes

These segments are outlined in figure 4.1 as part of an overall sport participation map. The definition of these segments and the nature of the sport participation map need to be defined based on the needs of each sport and country. This

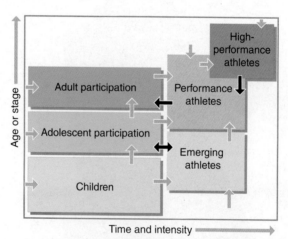

FIGURE 4.1   Sport participation spectrum.

Coaching Focus

> The needs and motives of people taking part in sport change at different stages of their lives. Coaches' philosophy, knowledge and capabilities need to reflect this in order to maximize the chances of athletes and participants having positive developmental experiences in and through sport.

approach may also involve the development of two maps: the existing position and the desired future position over specified timescales.

The sport participant segments are interconnected, as shown in figure 4.1. Individuals may enter or move through and between the various groups at different stages of their lives. Likewise, participants may move from one sport to another or partake in more than one sport simultaneously. Because there are so many variables, sporting pathways are individual, context specific and non-linear in nature.

Athlete Development

The participation segments have been established through the research and examination of existing athlete development models and front-line evidence. One of the most popular models describing phased engagement in sport, long-term athlete development (LTAD), was developed by Istvan Balyi[2] and is outlined in figure 4.2.

LTAD uses physical and psychological markers to propose a continuum of developmental stages and the associated coaching practices to maximize the performance potential of each individual and foster lifelong participation. While there has been questions raised about this model in the research literature, it has been adopted and adapted by several sports and countries. The model must

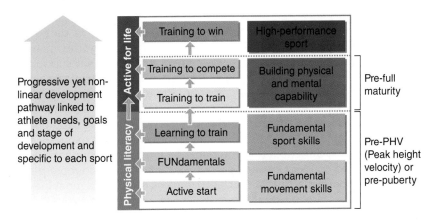

FIGURE 4.2 Long-term athlete development.

be adjusted to meet sport- and country-specific needs, including the objectives and titles of the stages and the associated age categories.

The difference between the real and ideal model in any sport and context should also be recognised, and interventions must be identified in order to close the gap between the two (coaching is one of these key interventions).

A complementary approach, the developmental model of sport participation (DMSP), identifies recreation and performance trajectories in sport. This model, originated by Jean Côté,[3] identifies three key phases in young people's sport participation. As with LTAD, the extent to which DMSP exists in different sports and countries will vary depending on culture, tradition, structures and available opportunities. Again, this may require adjustment of ages and stages to reflect the real and ideal scenario in any given sport or country. Following are the phases of the DMSP model:

- **Sampling phase.** Children take part in different activities and develop all-round foundational movement skills in an environment characterised by fun and enjoyment.
- **Specialising phase.** Children begin to focus on fewer sports, possibly favouring one in particular.
- **Investment phase.** Young athletes commit to achieving a high level of performance in a specific sport.

For the recreation trajectory, the sampling phase is instead followed by recreational years, in which children or adolescents continue to take part in sport for social interaction, healthy lifestyle and sheer enjoyment.

Contextual Fit

These models provide a basis for identifying key areas of focus in coaching at the various stages, taking account of individual needs and rates of development along the way.

The European Framework for the Recognition of Coaching Competence and Qualifications (EFRCCQ) established the classification of participation and performance coaching as the two main **coaching categories** making up the professional area of sport coaching. The current document supports and further refines this classification, reflecting the role of these two categories of coaching in responding to the participation and performance orientation of sport participants.[4]

Six **coaching domains** are proposed across the two coaching categories (see figure 4.3). The number and makeup of the coaching domains may vary among nations and sports, depending on the participation and performance pattern. Each sport should conduct an analysis of its participant base to more precisely determine the contextual fit and coaching domains required to meet the needs of participants. It is also important to recognise that coaches may work simultaneously across the two main categories and within different domains depending on the nature of their roles.

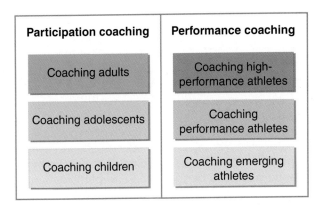

FIGURE 4.3 Coaching categories and coaching domains.

Research shows that the developmental needs and goals of athletes vary across the classifications and domains, and these should align with the capabilities required of coaches to fulfil them. Coach education and qualifications must account for both participation and performance coaching. A coach's developmental journey should reflect the domains in which they work, a point reinforced in chapter 5.

[1]Lyle, J. (2002). *Sport coaching concepts: A framework for coaches' behaviour.* London: Routledge.

[2]Balyi, I., & Hamilton, A. (1995). The concept of long-term athlete development. *Strength and Conditioning Coach,* 3(2), 5-6.

[3]Côté, J. (1999). The influence of the family in the development of talent in sport. *The Sport Psychologist,* 13, 395-417.

[4]EFRCCQ was developed by the European Coaching Council after a two-year review process. European Coaching Council. (2007). *Review of the EU 5-level structure for the recognition of coaching qualifications.* Koln: European Network of Sport Science, Education and Employment.

Coaching Roles

Coaches take on roles that entail varying levels of responsibility, competence, complexity and autonomy. Some coaches play roles that include coaching on a volunteer *and* part-time paid basis, coaching children *and* high-performance athletes and operating in both performance *and* participation coaching categories. For other coaches, status, coaching domain and coaching category are clearer (for example, a full-time paid coach who works with high-performance athletes or a volunteer coach who works with young children to support their early involvement in sport).

Just like athletes, coaches build up their expertise over time, and a significant part of their learning occurs on the job. Increased experience and capability are often accompanied by higher levels of responsibility and more complex roles.[1]

Clarification of coaching roles allows for the definition of core competences needed to fulfil them. This will, in turn, assist in charting coaches' education and broader development, employment pathways (in the case of paid coaches) and deployment pathways (in the case of volunteer coaches). Clear terminology on coaching roles also provides a basis for developing coach education programmes and qualifications that have a strong focus on the acquisition of job-related competences.

Such clarity will support training providers, be they federations or educational institutions in labelling and linking their qualifications (such as level 1, coaching diploma, or coaching certificate) to a common reference point that has clear application to the coaching workplace (for example, Senior Coach in the Coaching of Children). Descriptors of coaching roles and more clearly labelled coaching qualifications will also help employers identify suitable candidates and potential training needs.

Four main coaching role descriptors are proposed across both of the participation and performance coaching categories: Coaching Assistant, Coach, Advanced/Senior Coach and Master/Head Coach (see table 5.1).

Each of these four coaching roles consists of core functions, the nature of which will vary according to sport, country and context in which the coach is engaged. Due to considerable variances across cultural and sport contexts, the labels of Advanced/Senior Coach and Master/Head Coach merit further consideration.

The progression from Coaching Assistant to Master/Head Coach is hardly automatic. For many reasons, coaches may wish to remain in one of the three other roles specified. Some coaches, though extremely knowledgeable and experienced, prefer certain duties for which they may be or seem overqualified.

Role Requirements

These proposed classifications of coaching roles are tied directly to on-the-job activities of coaches. They do not necessarily reflect four levels of coaching qualifications. However, sport federations, coaching organisations and educational institutions are encouraged to align their educational and qualification systems with these roles.

Of course, completion of an educational qualification in no way guarantees that a coach is able to fully discharge the role-defined duties at a high level. Coaching courses that embed on-the-job training in which core competences must be demonstrated on a consistent basis provide greater assurance that the appropriate link between qualifications and role will be made.

A competence-based set of role descriptors, therefore, provides an alternative focus to traditional levelled qualifications that emphasize generic knowledge and are often not related to the requirements of the job. Such qualification systems have tended to associate lower levels of qualification with the coaching of children and young participants and the higher tiers with coaching in the performance context.

TABLE 5.1 Coaching Roles

Role descriptor	Knowledge and competence[1]
Coaching Assistant (*Note:* the term *Assistant Coach*, as opposed to the role of Coaching Assistant, may be applied at a number of levels. For example, an Assistant Coach in high performance might be operating at the level of the Advanced/Senior Coach).	Assists in the delivery of sessions[2]. Plans, delivers and reviews basic coaching sessions, sometimes under supervision. Basic level of knowledge, competence and decision making to deliver the primary functions with guidance. Supports the engagement of pre-coaches.
Coach	Plans, delivers and reviews coaching sessions over a season, and sometimes part of a wider programme. Extended level of knowledge, competence and decision making to independently deliver the primary functions. Supports the engagement and development of pre-coaches and Coaching Assistants.
Advanced/Senior Coach	Plans, delivers, leads and evaluates coaching sessions and seasons. Extended and integrated knowledge, competence and decision making to deliver the primary functions and to mentor others. Works independently and plays a leading role in the structure of the programme. Manages the development of Coaches, Coaching Assistants and pre-coaches.
Master/Head Coach	Oversees and contributes to the delivery, review and evaluation of programmes over seasons in medium- to large-scale contexts, underpinned by innovation and research. Specialist and integrated level of knowledge and competence, recognized as an expert with highly developed decision-making skills. Often involved in designing and overseeing management structures and development programmes for other coaches.

[1] See chapter 3 for detail on primary functions and chapter 6 for further detail on knowledge and competence.

[2] Sessions include both practices and competitions.

Staff Assignment and Synergy

These four descriptors of coaching roles may be applied to each coaching category and domain and tailored to the needs of each sport and country. In addition to the descriptors of coaching roles, several related roles are recognised. In high-performance coaching, for example, Advanced and Head Coaches frequently are responsible for other coaches and support personnel. Indeed, in some sports, coaches operating at the highest level of performance sport are called managers. In many cases they may have Assistant Head Coaches working alongside them (as opposed to Coaching Assistants). Increasingly, senior roles are also linked to performance management functions that may or may not be taken on by the coach.

> Though it is important for members of a coaching staff to relate well to one another, it is equally if not more important that together they provide the full complement of competences to ensure the proper instruction, support and development of the athletes under their direction.

A similar issue exists in the coaching of children, albeit in a very different context. There is a need for expert coaches to work with children because of the complexity of the developmental process and the multifaceted environment in which the coaching occurs. For example, many novice and volunteer coaches work for relatively short periods in coaching children, and there is a need to guide, coordinate, support and monitor this input. Thus, while the Head Children's Coach might not be a performance manager, he or she will play a key role in managing the environment, developmentally appropriate programmes and the fluid workforce that are often associated with the coaching of children. The process of applying coaching role descriptors to various contexts is outlined in figure 5.1.

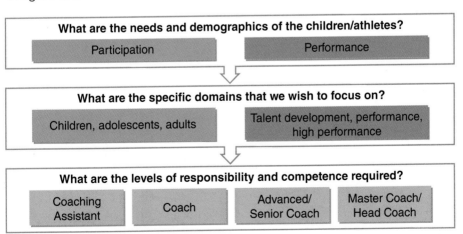

FIGURE 5.1   Defining coaching roles for sport, country and organisation.

This approach will provide the basis for a participant-centred assignment of coaching staff and for the induction of new coaches and those in pre-coaching roles. It will also assist in promoting synergies between coaches working in different roles and domains in the overall coaching structure.

[1] Côté, J., Erickson, K., & Duffy, P. (2013). Developing the expert performance coach. In D. Farrow, J. Baker, & C. MacMahon (Eds.), *Developing elite sport performance: Lessons from theory and practice* (pp. 17-28; 2nd ed.). New York: Routledge.

6

Coaching Knowledge and Competence

To deliver effective programmes to meet athletes' needs and deliver sustained, guided improvement, coaches must develop knowledge and competence in a range of areas. This can be accomplished by gaining additional experience, learning on the job, engaging in formal and informal education and networking with other coaches.

Knowledge Areas

Jean Côté and Wade Gilbert (2009, p. 316)[1] have described effective coaching in the following way:

> The consistent application of integrated professional, interpersonal, and intrapersonal knowledge to improve athletes' competence, confidence, connection, and character in specific coaching contexts.

In this definition, three areas of knowledge are central. Gilbert and Côté (2013)[2] have further defined these knowledge areas: professional knowledge (content knowledge and how to teach it), interpersonal knowledge (relates to the ability to connect with people and is closely related to emotional intelligence) and intrapersonal knowledge (of self, based on experience, self-awareness and reflection). The centrality of intrapersonal knowledge is supported by ICCE who advocate that coach behaviour should be underpinned by a clear set of values, coaching philosophy, ethical principles and responsibilities. Table 6.1 contains the elements of the three knowledge areas.

TABLE 6.1 Knowledge Areas

Professional knowledge	The sport
	Athletes
	Sport science
	Coaching theory and methodology
	Foundational skills
Interpersonal knowledge	Social context
	Relationships
Intrapersonal knowledge	Coaching philosophy
	Lifelong learning

Adapted from Côté and Gilbert (2009) and Gilbert and Côté (2013).

Competences

Each of these areas of knowledge underpins the competence of the coach to do the job across the primary functions outlined in chapter 3. Coaching competences can be delineated into those that permit one to meet the needs of a specific situation (functional competence) and those that enable one to perform specific duties (task-related competence).

> **To be effective in their roles, coaches need to develop competences in each of the six primary functions that are underpinned by professional, interpersonal and intrapersonal knowledge.**

The relationship between coaching values, knowledge and functional competence across the coaching strands and domains is outlined figure 6.1.

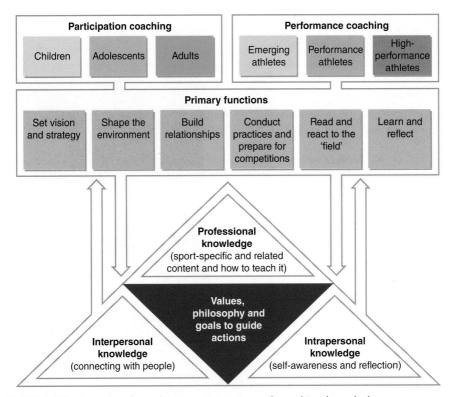

FIGURE 6.1 Functional coaching competence and coaching knowledge.

Functional Competence

Functional coaching refers to adopting an approach to guide the improvement of athletes in a given social and organisational context. It acknowledges that coaching is a complex and dynamic activity that spans beyond the track or the pitch and the mere transfer of knowledge and skills from coach to athlete. Essentially, coaches must be equipped to understand, interact with and shape their environments. They must be able to demonstrate task-related competence in each of the primary functions identified in chapter 3.

Task-Related Competence

Coaches carry out a range of tasks that require a variety of competences. These can be classified according to the six primary functional areas as outlined in table 6.2. Whilst coaches can develop their competency on the job, the task-related competences should be embedded in basic qualifications for coaches. Table 6.2 indicates the match of levels of competency with coaching roles.

TABLE 6.2 Levels of Competence and Responsibility in Coaching

Functions	Competences	Coaching Assistant	Coach	Advanced/ Senior Coach	Master/ Head Coach
Set the vision and strategy	Understand big picture				
	Align and govern				
	Analyse needs				
	Set the vision				
	Develop strategy				
Shape the environment	Create action plan				
	Organise setting and personnel				
	Identify and recruit athletes, staff and resources				
	Safeguard participants				
	Develop progress markers				
Build relationships	Lead and influence				
	Manage				
	Manage relationships				
	Be an educator				

Functions	Competences	Coaching Assistant	Coach	Advanced/ Senior Coach	Master/ Head Coach
Conduct practices and structure competitions	Guide practice				
	Employ suitable pedagogy or andragogy				
	Identify and manage suitable competitions				
Read and react to the field	Observe				
	Make decisions and adjust				
	Record and evaluate				
Learn and reflect	Evaluate session and programme				
	Self-reflect and self-monitor				
	Engage in professional development				
	Innovate				

Colors denote levels of competency on a scale lowest (white) to highest (red).

Ongoing research is attempting to identify specific learning outcomes related to coaches' knowledge, competence and values.[3] The alignment of qualifications among national federations, government organisations and international federations provide options, pathways and recognition for coaches. The education, development and certification of coaches are addressed in chapters 8 and 9.

---

[1] Côté, J., & Gilbert, W. (2009). An integrative definition of coaching effectiveness and expertise. *International Journal of Sports Science and Coaching*, 4(3): 307–323.

[2] Gilbert, W., & Côté, J. (2013). Defining coaching effectiveness: a focus on coaches' knowledge. In W. Gilbert (Ed.), *Handbook of sports coaching*. London: Routledge.

[3] See Abraham, A., & Collins, D. (2011). Taking the next step: Ways forward for coaching. *Quest* (63), 366-384; Schempp, P.G., & McCullick, B. (2010). 'Coaches' expertise' in J. Lyle and C. Cushion (Eds.), *Sports coaching: Professionalisation and practice*. Edinburgh: Churchill Livingstone Elsevier; as well as Gilbert, W. & Côté, J. (2013), as outlined in note 2.

7
Coaching Objectives

Assuming a coach has the knowledge and competence for the role, it's time to establish more specific objectives. These goals should stem from positive values and a desire to create the best possible experience for the athletes. Coaches should reflect on and define their objectives in their own words, taking full account of the programme aims in which their work occurs. To assist in this effort, many classifications of athlete outcomes and coaching objectives have been developed by practitioners, federations and researchers over the years.

Developing the Whole Athlete

Classifications focus on the sport-related competences, such as technical, tactical, physical and mental. In other cases, the emphasis is on the outcomes relating to the holistic development of the participant, such as the 5 Cs: competence, confidence, connection, character and caring.[1] Either way, coaches should see the athlete as a whole person with individual needs and preferences.

Current research is also seeking to establish a more comprehensive view of the specific outcomes associated with children's sport.[2] In the South African context, the outcomes associated with effective coaching in the children's domain have been grouped into three main categories of competence:[3]

1. **Sport competences.** Technical, tactical and physical capabilities required for participation at various levels. These competences form the traditional core of sport and occur in the context where participants strive for and deal with the consequences of competition, success and failure.

2. **Personal competences.** Capabilities that relate to the development of the whole person and may be supported and developed through participation in sport. These have been grouped into social, cognitive and emotional outcomes.

3. **Life course competences.** The combination of sport and personal competences and experiences that positively contribute to the individual life course of the participant. (The ability to apply effort to undertake practice and achieve a goal can be applied to study in the school and college context, or the development of a sporting talent leading to a professional career as a player or coach, or the adoption of a lifestyle that promotes a positive approach to health and fitness and daily living.)

> **Coaches are responsible for identifying the specific objectives they are seeking to achieve with their athletes. Coaches should also develop a clear sense of why they are striving for these objectives, informed by their values, philosophy and interaction with athletes.**

Teaching Lifelong Lessons

Based on experiences, working context and access to research and good practice, each coach will develop a unique way of describing personal coaching objectives. This description should include a healthy balance of sport, personal and life course outcomes. This balance will take different shapes depending on the needs, ages and stages of development of the athletes. Coaches should keep the following considerations in mind as they describe the desired outcomes of their work with athletes:

- **Respect and opportunity.** Participants should learn to respect the sport, administrators, officials, coaches, teammates and opponents and see them as partners with the same objectives and goals. In return, they should feel that they are being provided with access to valuable opportunities in sport.

- **Fair play and sportsmanship.** Through competitive sport experiences, participants can learn what behaviours are appropriate, the functional and relational benefits of such behaviours and the consequences of acting inappropriately.

- **Trust and teamwork.** Children can learn the value of working as part of a group and the importance of reciprocal trust for the achievement of mutual goals, whilst older athletes should demonstrate these attributes on a consistent basis.
- **Health and fitness.** With appropriate fitness training programmes through the developmental years, sport should set participants on the right path for a lifetime of physical activity and health.
- **Competition and success.** Sport participants should learn how to compete. Some will take to it easily, but others will need guidance in respecting the rules and opponents as well as in dealing with both victory and defeat. Through effective reinforcement, athletes should come to appreciate the link between effort and personal improvement, regardless of the win–loss record.
- **Fun and lifelong engagement.** Coaches have a responsibility to ensure that young people and athletes, while challenged, enjoy their participation and are intrinsically motivated. This is the first step in fostering lifelong participation. By stating their objectives in terms of overall outcomes that account for participants' needs and stages of development, coaches will be more effectively positioned to guide improvement. This will also help them to reflect on the way in which they conduct their coaching sessions and interactions.

The outcomes and objectives outlined in this chapter offer options for coaches, which assist in the design of coach education programmes while encouraging each coach to develop personal values, philosophies and objectives.

[1] These outcomes were first identified by Richard Lerner's group in the wider youth development context. Lerner, R.M., Lerner, J.V., Almerigi, J.B., Theokas, C., Phelps, E., Gestdottir, S., Naudeau, S., Jelicic, H., Alberts, A., Ma, L., Smith, L.M., Bobek, D.L., Richman-Raphael, D., Simpson, I., DiDenti Christiansen, E., & von Eye, A., (2005). Youth development contributions of fifth-grade adolescents: Findings From the first wave of the 4-H study of positive. *The Journal of Early Adolescence, 25* (1) 17-71. These have been refined to the sport context by Jean Côté et al. (Vierimaa, M., Erickson, K., Côté, J., & Gilbert, W. (2012). Positive youth development: A measurement framework for sport. *International Journal of Sport Coaching and Science,* 7 (3) 601-614).

[2] This research is being conducted by Lara-Bercial, S., Côté, J., Bruner, M., & Duffy, P.

[3] South African Sports Confederation and Olympic Committee (in press). *The South African Coaching Children Curriculum.* Johannesburg: Author.

8

Coach Development

Through the first seven chapters, two points have been emphasized: A coach's primary mission is to help athletes develop and improve, and to fulfil that aim coaches require functional competences that are informed by knowledge and reflection. So it stands to reason that coach education and development must support the establishment of effective behaviours, skills and attitudes and not merely the accumulation of knowledge.[1]

In turn, coaches should seek to enhance their abilities through a range of methodologies from directed, instruction-based approaches to more facilitative, collaborative means of learning, which includes interaction with other coaches. The most effective means of delivery meets the learner's (coach's) unique needs in the specific situation.

Courtesy of Professor Patrick Duffy and Leeds Metropolitan University

Long-Term Process

Becoming a fully educated and developed coach requires sufficient time, motivation, application and practice. Like athletes, coaches grow through exposure to learning situations and opportunities over a sustained period. The most common of these is on-the-job experience, often preceded by participation as an athlete. The gradual accumulation of these experiences contributes to a process of long-term coach development (LTCD).[2]

National and international federations have a key role in identifying the optimal developmental processes that will enhance the learning and progression of their coaches. Recent research has indicated that the combination of athletic experience, coaching experience and informal and formal education provides the basis that may lead to Advanced/Senior Coach roles in performance coaching (see figure 8.1).[3] The research found that coaching certification was not always associated with the developmental pathways of performance coaches. While in many cases this was somewhat compensated for by other forms of formal and informal education, the research findings highlight the need for greater emphasis on and accessibility to formal, certified coach education programmes.

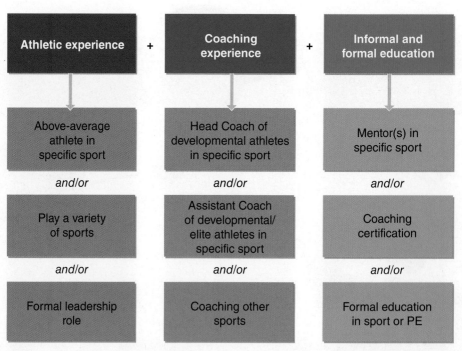

FIGURE 8.1 Experiences contributing to the development of performance coaches.
Adapted from Côté, Erickson & Duffy, 2013.

Educational Curriculum

Each sport and nation should develop a clear view of what LTCD means in their context. This will provide a firm foundation for curriculum development and delivery that includes a variety of learning experiences to meet the needs of coaches at each stage of development.

In support of the curriculum development process, the present *Framework* proposes a classification of learning situations adopted from the work of Jennifer Moon. This classification distinguishes between two main types of learning situations: mediated and unmediated (see figure 8.2).

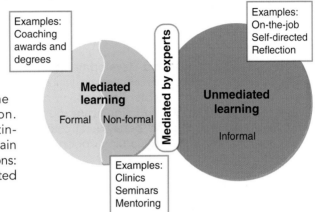

FIGURE 8.2 Types of learning situations for coaches.

Much of **unmediated learning**, according to Moon, refers to when coaches initiate their learning, choose what they want to learn and decide how they wish to learn it. Examples of unmediated learning are reading a book, watching a DVD and reflecting on prior experiences.

Mediated learning refers to learning that is aided directly by another person or through the use of a medium that simplifies the material of teaching. Mediated learning is achieved through two modes of education: formal and non-formal:

- Formal education "takes place in an institutionalized, chronologically graded, and hierarchically structured educational system."[4]
- Non-formal education is any "systematic educational activity conducted outside the framework of the formal system to provide select types of learning to particular subgroups in the population."[5] In coaching, the context might include clinics, seminars and mentoring.

Research[6] suggests that coaches learn best when

- their prior experiences and abilities are recognised and they are encouraged to reflect and build on them;
- they are motivated to learn and find the relevant learning materials;
- they are encouraged to take responsibility for their learning;

- the climate is positive and supportive to minimise anxiety, encourages experimentation and challenges them appropriately;
- educators take account of the way they like to learn;
- they have plenty of opportunities to practise and apply the information to their own context;
- they are involved and engaged in their own learning; and
- they experience some success and gain feedback that builds their self-confidence.

Coaches' educational and developmental experiences should mirror the complex and changing environments in which they operate. Effective coach development should offer a blended learning package of mediated and unmediated learning situations and significant job-related components. Together these promote learning and behavioural change and encourage coaches to seek additional self-directed learning opportunities.

Experiential Learning and Mentorship

Traditionally, coach education has provided mediated, formal learning situations. Very often, coach education, both formal and non-formal, is classroom based, assessment focused and qualification driven. However, coaches learn well from practical experience and interaction with other coaches,[7] highlighting the need to balance formal coach education with learning experiences on the field, court and track and in the pool and gym.

Indeed, **experiential learning** is central to coach development. However, it is very different to learning from experience. It is intentional and can be mediated or unmediated. Through exposure to a broader range of situations, experiential learning provides coaches with a chance to discover what knowledge and skills they already have and to enhance their decision-making capabilities across a broader spectrum of coaching circumstances.

Experiential learning requires self-awareness and self-reflection on the part of the coach. Awareness and reflection are teachable skills and should be a central part of a balanced coach development programme.

An effective on-the-job learning option is integrating trained mentors and communities of practice[8] into coaching delivery. Coaching organisations must also value, recognise, respect, trust and encourage the contribution that experienced coaches can offer in guiding, educating, mentoring and developing less seasoned colleagues. More senior coaches must also recognise that this mentoring and support role is a core part of their professional responsibility.

Coaches also gain knowledge from experts in related fields such as strength and conditioning, biomechanics, sport psychology and nutrition. The value of these exchanges increases when the knowledge is conveyed systematically, gauged at the appropriate level of complexity and combined with examples and applications to the coaching of the athletes.

Coach Development

> The process of coach development takes place over time and includes athletic experience, coaching experience, informal and formal and non-formal education. The design of coach education programmes should provide for learning opportunities that take account of coaches' experience and working contexts.

Delivery by Coach Developers

Federations, coaching organisations and educational institutions seeking to develop coaches are advised to consider how they identify and train **coach developers** who will deliver their coach education programmes. Coach developers must have a genuine passion for their tasks. They must be carefully selected and recruited, have a suitable support system and be evaluated regularly so that their competence and growth in the role can be assessed.

Coach developers also play a vital role in the delivery of non-formal learning situations and the promotion of coaches' engagement in unmediated on-the-job learning. Coaches with sufficient experience and a desire to develop other coaches' skills may wish to train to become coach developers.

Each sport and nation should establish a long-term career pathway for coach developers, clearly linked with and aligned to coaching categories, domains and roles. Figure 8.3 provides an overview of this pathway.

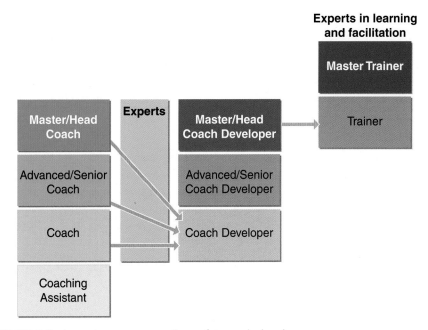

FIGURE 8.3 Long-term career pathway for coach developers.

Advanced/Senior and Master/Head Coach developers have an inherent responsibility in supporting the development of less experienced developers. Coach developer trainers (CDTs) play a significant leadership role in facilitating learning and mentoring.

The following steps will help to develop a sustainable, high-quality coach developer programme:

1. Analyse current and forecasted coach developers (numbers, skills and experience), taking into account the needs of the coaching workforce and the availability of more experienced coaches to support the development of other coaches.
2. Based on step 1, formulate a strategic plan.
3. Using established criteria, select from coach developer applicants.
4. In accordance with predetermined guidelines, induct, train, develop and support coach developers through the pathway.

Federations, countries and educational institutions should invest time and resources in creating an effective coach developer workforce. This will enhance their ability to develop existing coaches as well as increase the number of new coaches coming into the system in those areas where they are needed.

---

[1] Moon, J.A. (2004). *A handbook of reflective and experiential learning: Theory and practice*. London: Routledge-Falmer.

[2] South African Sports Confederation and Olympic Committee. (2012). *The South African model of long-term coach development*. Johannesburg: Author.

[3] Côté, J., Erickson, K., & Duffy, P. (2013). Developing the expert performance coach. In D. Farrow, J. Baker, & C. MacMahon (Eds.), *Developing elite sport performance: Lessons from theory and practice* (pp. 17-28; 2nd ed.). New York: Routledge.

[4] Coombs, P., & Ahmed, M. (1974). *Attacking rural poverty: How non-formal education can help*. Baltimore: Johns Hopkins Press, 8.

[5] As note 3.

[6] Knowles, M.S., Holton, E.F., & Swanson, R.A. (1998). *The adult learner* (5th ed.). Woburn, MA: Butterworth-Heninemann.

[7] Carter, A., & Bloom, G. (2009). Coaching knowledge and success: Going beyond athletic experiences. *Journal of Sport Behavior*, 32 (4), 419-437.

[8] Lave, J., & Wenger, E. (1998). Situated learning: Legitimate peripheral participation. In Wenger E., *Communities of practice*. New York: Cambridge University Press.

9

Coach Certification and Recognition

Coaching roles, competences and qualifications should be closely correlated. When these three key facets are not aligned, the result can range from disappointment to full-blown tragedy. For example, a coaching assistant who played sport at an elite level but had no education or training in teaching age-appropriate skills would likely miss the mark in technical instruction and discourage even the most eager learners. Therefore, the first point of emphasis is to ensure that all coaches can fulfil their basic duties.

Many coaches are well qualified for their roles, and federations, countries and educational institutions increasingly appreciate the importance of effective and relevant coach development. This trend is encouraging, because the potential harm done by one who is unprepared or otherwise incapable of fulfilling the responsibility of a coaching position is reason alone to insist on preparatory measures.

Courtesy of Richard Juilliart

On May 3, 2013, in Lausanne, Switzerland, 28 coaches successfully completed the International Coaching Enrichment Certification Program.

Educational Requirements

Depending on their quality and rigour, coach education, certification and licensing programmes can ensure that all participating coaches have an acceptable level of competence. A certificate or licence from an approved coach education programme ensures quality in the coach development process. Well-structured educational coursework and on-the-job mastery meriting the award of a certificate or licence benefit everyone in sport.

It is, therefore, recommended that national and international federations carefully analyse and determine the mandatory certification and licensing requirements for designated roles in their sports. This analysis should take account of relevant national and international qualification frameworks and any legislative requirements that exist or are envisaged.

Coach certification varies according to sport, nation and institution. Federations at the national and international levels take the lead in sport-specific certification courses. Institutions of higher education tend to focus on more universal topics such as sport science, coaching methods and theory. There also is a positive trend for such institutions to include sport-specific components. It is recommended that partnerships be fostered between federations and educational institutions to maximize the quality and relevance of the courses offered to coaches. Figure 9.1 outlines a process to maximize this partnering approach to curriculum design and coach certification.

FIGURE 9.1　Curriculum design process.

Qualifying Standards

Coaches devote much time and energy to developing athletes. The majority of coaches are volunteers, yet they belong to the same blended profession as part-time paid and full-time paid coaches.

National and international federations should specify the standards for particular coaching roles. Coaches in all four roles—Master/Head Coach, Advanced/Senior Coach, Coach, and Coaching Assistant—should be certified for their organisation, sport and nation. This certification should provide evidence of how the programme has prepared the coach for functional competence in a specific role and coaching domain. Some certification programmes will be designed in such a way that access to more than one coaching domain is provided for gaining additional expertise. Regardless of the structure of the programme, coach certification should clearly indicate the coaching roles and coaching domains for which the coaches have been prepared.

In the case of pre-coaches, their roles need to be clearly defined and the time scales and expectations for progression to formal coaching roles and the associated certification should be defined.

> Coaches should be appropriately trained and qualified for the roles that they play and the domains in which they will work. National and international federations should specify the designations (certification and licensing) required for specified roles, taking account of relevant national and international qualification frameworks.

Awards and Designations

Coaches deserve to receive appropriate recognition nationally and internationally for their expertise and qualifications. Coaching qualifications should be referenced against appropriate national and international standards and benchmarks in educational and vocational training.

Figures 9.2 and 9.3[1] present a model showing how certification programmes of national and international federations might align with coaching roles and with other forms of educational advancement. An increasing number of countries

Coaching roles	Achievement standards		
	National and international federation levels	University/ higher education awards	Other coach education institution and agency awards
Master/Head Coach	Level 4	University degree or postgraduate degree	
Advanced/ Senior Coach	Level 3	University diploma or degree	
Coach	Level 2		Coaching certificate
Coaching Assistant	Level 1		Coach introductory course award

FIGURE 9.2 Alignment of achievement standards with coaching roles.

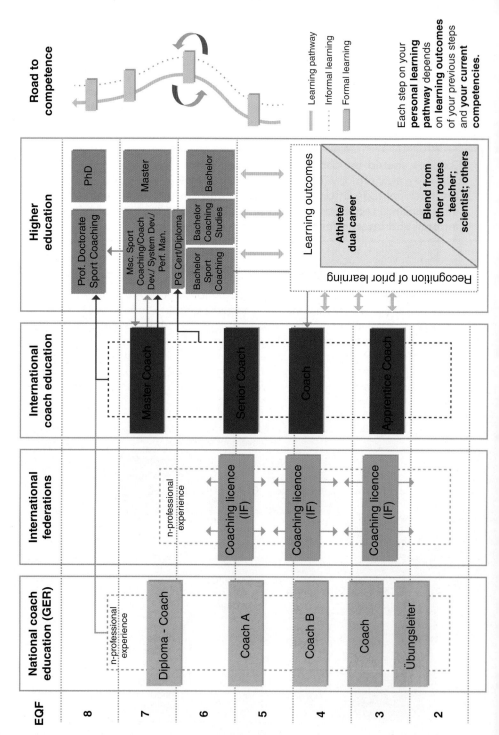

FIGURE 9.3 Long-term learning pathway for coaches.

may require coaching preparation, certification and continuing education programmes to align with national qualification frameworks.

Formal coaching certification systems should also factor in the **recognition of prior learning** and experience (RPL) of coaches.[2] This is particularly important given the demonstrated benefits of experiential learning in coaching. Also, in sports and countries with relatively new certification programmes, recognition of previous coaching experience is a means of opening the pathway to higher levels in the system until those seeking to advance complete their coursework.

As the relationship between formal coaching curricula and on-the-job training is clarified, systems for licensing coaches can be introduced. Coaching licences, which attest to competency and right to practise, are particularly relevant for all coaches earning income from their work and for volunteers operating in Master/Head and Advanced/Senior Coach roles.

All of these measures will lead to even better developmental, educational and vocational experiences for coaches. The recognition of coaching qualifications and the mobility of coaches will also be enhanced as part of a blended profession across the globe.

---

[1] This overview has been adapted from ongoing work between Leeds Metropolitan University, ICCE and Trainerakademie (Koln, Germany) on postgraduate-level pathways for coaches and coach developers and on the creation of an international professional doctorate in sport coaching. This work has been informed by earlier research into coaching pathways in higher education as part of the wider AEHESIS project: Duffy, P. (2007). Coaching: Final report. In Froberg, K; Petry K. (Eds.), *Aligning a European Higher Education Structure in Sport Science (AEHESIS) Final Report.* Cologne: German Sports University.

[2] The South African Qualifications Authority (SAQA) and the South African Sports Confederation and Olympic Committee (SASCOC) have agreed to the establishment of a National Pilot RPL programme for coaches and coach developers as part of the implementation of the South African Coaching Framework.

Coaching Framework Applications

The *International Sport Coaching Framework* is an important step in supporting the creation of relevant, sustainable and high-quality coach education, development and deployment systems worldwide. Adaptation and implementation of the principles contained in the *ISCF* to sport- and country-specific circumstances offer multiple benefits.

Create High-Quality Coach Education and Development Programmes

The ultimate objective is to support the effort to further enhance the quality of coaching in different sports and countries. The *Framework* is not mandatory, but it presents a series of practice- and research-supported principles and tools that may be applied to specific contexts. Coach education and development programmes that follow the recommendations will bolster coaches' competence, which will benefit athletes at all levels.

The *ISCF* provides a transparent tool that assists in the design of new programmes and benchmarking or aligning existing programmes (see figure 10.1).

This tool can be used in a number of circumstances:

- Coach education and development programme providers can compare their own qualifications with an internationally recognised framework regarding learning outcomes. This may be particularly relevant in the connection between the systems that operate worldwide as well as between federation and higher education qualifications.

- Coaching organisations and those employing and deploying coaches can more effectively assess the competences of coaches coming from different sports or nations and in specific roles. This will assist in the recognition of coaching qualifications and prior learning as well as in the identification of gaps in competency.

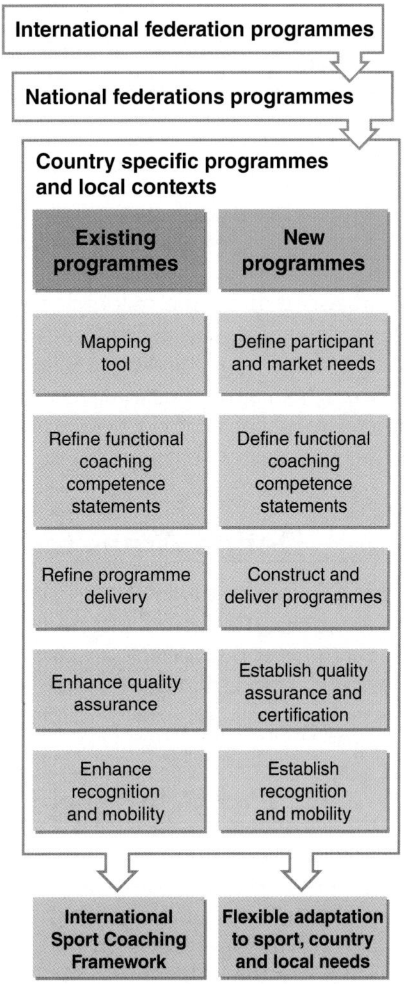

- Nations or federations looking to develop new systems or qualifications can use the *ISCF* to help determine standards of coach qualification for each level and the necessary content to fulfil the associated requirements.

- Those who train and employ coaches can use the competences defined in the *ISCF* as an assessment and development tool for their coaches, leading to the identification of training needs.

FIGURE 10.1 Applications of the *International Sport Coaching Framework*.

> The application of the *Framework* to specific contexts will enhance the quality of coach education and development in existing and new programmes. It will enhance political and legal decision making and promote systemic and progressive improvement and quality assurance in coaching.

Evaluate and Improve Existing Programmes

The *ISCF* provides a prism through which to identify, implement and evaluate practices in coach education and development. It highlights the building blocks that affect quality, efficiency and effectiveness and provides a set of tools and concepts that can be tailored to specific contexts.

By analysing the *ISCF* blueprint and comparing it to their own programme design, sport administrators may be better able to prioritise and allocate resources in a manner that maximizes their return on investment. Quality assurance processes applied over time will underpin the continuous improvement of coaching and coach development on a systemic and sustainable basis.[1]

Define Areas for Research and Evaluation

The *ISCF* offers clarity regarding both the desired competences of coaches to maximize participation and performance as well as the key components of systems that support coach education and development. This creates a potential index of areas for future research and evaluation, which will advance the blended profession towards new degrees of expertise and effectiveness.

Consider and Make Political Decisions

The *Framework* can also serve as a basis for creating, evaluating and revising regulations and laws to underpin the quality, sustainability and blended professional nature of sport coaching. This will allow for more uniform and measured political and legal decision making by administrators, boards, and authorities throughout the sport and educational communities.

Stimulate Global Exchange

The establishment of a common language will facilitate the exchange of information and knowledge between partners, and even competitors, within and across countries and sports. This will in turn enhance understanding between them at all levels and open new avenues for cooperation as well as clear templates for the recognition of coaching qualifications between countries. Members of the sport community who stand to benefit from the *ISCF* are shown in figure 10.2.

FIGURE 10.2 Potential beneficiaries of the *Framework*.

Promote Further Refinement

The response to Version 1.1 signalled many potential outgrowths of the *ISCF*, some of which have already made an impact.

Since the release of Version 1.1 at the Global Coaches House in London during the 2012 Olympic and Paralympic Games, the *ISCF* has provided a platform for further research and consultation. Through open dialogue, the Project Group has further refined the *Framework* to produce version 1.2 for the Global Coach Conference in South Africa in September 2013.

Further refinement of the *ISCF* is also tied to the new *International Sport Coaching Journal* (ISCJ), a publication of the ICCE and the American Alliance for Health, Physical Education, Recreation and Dance, which will frequently include research and commentary on the *Framework*. Through this ongoing critical examination, *ISCF* will be further refined for the timely release of Version 2.1 at the Global Coaches' House in Rio de Janeiro at the Olympic and Paralympic Games in 2016 (see figure 10.3).

FIGURE 10.3 *Framework* development process.

In valuing the work of coaches and advocating an athlete-centred focus, *ISCF* signposts actions that will enhance the sport experience for all. Sport organisations, coaches and athletes will be the most apparent beneficiaries, but the positive ramifications of this and subsequent iterations of the *Framework* will be felt worldwide.

[1] In parallel with the process to develop the *International Sport Coaching Framework*, a Quality in Coaching innovation group of leading agencies within ICCE has developed a Draft Quality in Coaching Model, which is undergoing pilot testing and research.

Glossary

athlete-centred coaching—Coaching based on recognising the needs of the athlete, planning and delivering a practice and competition programme accordingly. Coaching should be informed by an understanding of the process of long-term athlete development (LTAD).

coach developer, educator, or tutor—Formal roles that entail developing other coaches through mediated learning and encouraging coach engagement in unmediated learning.

coach development—The range of experiences and learning opportunities available to coaches that support the development of their expertise in fulfilling specified roles. Over a coach's career, this process might be described as long-term coach development (LTCD).

coaching categories—The division of the main focus of coaching into participation and performance, recognising that some people may coach in both categories.

coaching domain—Areas of coaching that are identified through the types of participant coached (i.e., children; adolescent and adult participants; and emerging, performance and high-performance athletes).

coaching role—The specific job carried out by the coach that is defined by employers (for paid coaches) and deployers (for unpaid coaches).

coaching status—The classification of coaches according to their volunteer, part-time paid or full-time paid input to athlete development. Pre-coaching roles are also included in the classification of coaching status.

coaching system—The structures and delivery mechanisms in any given sport or nation to support coaches and the development of coaching.

competence based—Formal assessment based on the demonstration of required competence in fulfilling the coaching role.

entourage—Used by the International Olympic Committee to refer to the range of individuals and roles that are associated with an athlete's involvement in sport, such as parents, coaches, sport scientists, managers, and performance managers. The term, while developed for high-performance sport, also has meaning for youth participating in sport.

experiential learning—Learning that occurs through the resolution of real-life problems. It is structured and managed and not the same as learning from experience.

mediated learning—Learning that is aided directly by another person or a medium that simplifies the material (e.g., qualification, online course, clinic).

pre-coaching role—The function of parents, players or volunteers in supporting the work of the coach. This role does not require a formal qualification but should be supported through guidance and training and provide a pathway into more formal coaching roles and qualifications.

participation sport—Sport in which the main priority is to achieve self-referenced outcomes such as personal improvement, healthy lifestyle, socialisation or enjoyment.

performance sport—Sport in which the development and demonstration of capabilities referenced against normative standards in competition are the main objectives.

recognition of prior learning—Assessment of a coach's competence that takes into account previous experiences and learning of the coach (even in other fields).

unmediated learning—Learning that is initiated, driven and sustained by the learner without any external facilitation (e.g., a book, self-reflection, watching a competition). Also known as informal learning.

About the Authors

The *International Sport Coaching Framework* is a joint endeavour led by the International Council for Coaching Excellence (ICCE) and the Association of Summer Olympic International Federations (ASOIF), supported by Leeds Metropolitan University (LMU) through project administration, technical advice and research. The editors of the publication are Patrick Duffy (professor of sport coaching at LMU and vice president of ICCE), Mark Harrington (general manager-technical services of the International Rugby Board and chair of the Development and Education Group of ASOIF) and Sergio Lara Bercial (senior research fellow at LMU and ICCE technical officer).

International Council for Coaching Excellence

Formerly the International Council for Coach Education, the ICCE was established in September 1997 as a not-for-profit international organisation with the aim of promoting coaching as an internationally accepted profession. ICCE members seek to enhance the quality of coaching at every level of sport.

More specifically, the ICCE's mission is to lead and support the global development of coaching as a blended profession and to enhance the quality of coaching at every level in sport, guided by the needs of members, federations, nations and key partners.

The ICCE's strategic objectives are to fortify its organisational infrastructure, develop an international sport coaching framework, build a community of coaches globally and strengthen the position of coaching as a profession. ICCE partners and markets include national representative bodies responsible for coach development, international federations, institutions that deliver coach education or represent coaches, individuals who design and deliver coach education, coaches and the international sport community at large. Visit the website at www.icce.ws.

Association of Summer Olympic International Federations

On May 30, 1983, the 21 international federations governing the sports of the 1984 Summer Olympic Games decided to form the Association of Summer Olympic International Federations. This alliance sought to address the issues of common interest in the Summer Olympic Games and the Olympic Movement and any other matter deemed necessary by the international federations.

More formally, ASOIF's mission today is to unite, promote and support the international summer Olympic federations and to preserve their autonomy while coordinating their common interests and goals.

The international federations have the responsibility to manage and monitor the daily functioning of the world's various sport disciplines, including the practical organisation of events during the Games and the supervision of the development of athletes practising these sports at every level. Each international federation governs its sport throughout the world and ensures its promotion and development. ASOIF's members now total 28. Visit the website at www.asoif.com.

Leeds Metropolitan University

Through its Carnegie Faculty, LMU has a long tradition in the professional preparation of graduates in physical education, sport science, sport development, physical activity and sport coaching. Through its Research Institute for Sport, Physical Activity and Leisure, the university plays an active role in research and enterprise in the UK and internationally. Since 2011, LMU has become the home of ICCE, and the Global Coaching Office is now housed in Headingley Carnegie Stadium. Visit the website at www.leedsmet.ac.uk.

Insightful new journal advances the profession of coaching

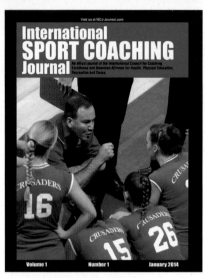

International Sport Coaching Journal
Wade Gilbert, PhD, Editor • Pat Duffy, PhD, Editor
Frequency: 3x per year (January, May, September)
Current volume: 1 (2014)
ISSN: 2328-9198 • **ISBN:** 978-1-4504-6879-4

Editors-In-Chief
Wade Gilbert
California State University-Fresno, USA
Pat Duffy (Interim Editor)
Leeds Metropolitan University, England

Associate Editor
Mike Sheridan, Ohio University, USA

Editorial Board
Kim Bodey, Indiana State University, USA
Jean Côté, Queen's University, Canada
Dan Gould, Michigan State University, USA
Koon Teck Koh, National Institute of Education, Singapore
John Lyle, Leeds Metropolitan University, England
Cliff Mallett, University of Queensland, Australia
Lutz Nordmann, Trainerakademie, Germany
Pierre Trudel, University of Ottawa, Canada
Penny Werthner, University of Calgary, Canada
Bingshu Zhong, Capital University of Physical Education and Sports, Republic of China

The *International Sport Coaching Journal (ISCJ)* seeks to advance the profession of coaching through research articles, informative essays, experiential accounts, and systematic applications that enhance the education, development of knowledge, leadership, and best practices of coaches.

A joint venture of the International Council for Coaching Excellence (ICCE) and American Alliance for Health, Physical Education, Recreation and Dance (AAHPERD), *ISCJ* will publish a blend of relevant studies, technical insights, examples of coaching methods employed around the world, engaging front-line stories, and thought-provoking commentaries.

The journal features scientific articles about coaching and coaching education that appeal to practicing coaches, administrators, and researchers; showcase best practices; and establish a more universal language in coaching around the globe.

The editorial board consists of 10 worldwide recognized experts on coaching, coaching research, and coach development. The quality, authority, and applicability of *ISCJ* makes it an essential resource and reference for all who have an interest in coaching and a stake in its improved development as a profession.

For more information or to order,
visit **www.ISCJ-Journal.com** or call:
(800) 747-4457 US • (800) 465-7301 CDN
44 (0) 113-255-5665 UK • (08) 8372-0999 AUS
0800 222 062 NZ • (217) 351-5076 International

ISCJ is now accepting submissions for manuscripts;
visit **http://mc.manuscriptcentral.com/hk_iscj**.